MW00876458

This Calendar Belongs To

_____

2022

January

Su	Mo	Tu	We	Th	Fr	Sa
26	27	28	29	30	31	1
2	3	4	5	6	7	8
9	10	11	12	13	14	15
16	17	18	19	20	21	22
23	24	25	26	27	28	29
30	31	1	2	3	4	5

February

Su	Mo	Tu	We	Th	Fr	Sa
30	31	1	2	3	4	5
6	7	8	9	10	11	12
13	14	15	16	17	18	19
20	21	22	23	24	25	26
27	28	1	2	3	4	5

March

Su	Mo	Tu	We	Th	Fr
27	28	1	2	3	4
6	7	8	9	10	11
13	14	15	16	17	18
20	21	22	23	24	25
27	28	29	30	31	1

April

Su	Mo	Tu	We	Th	Fr	Sa
27	28	29	30	31	1	2
3	4	5	6	7	8	9
10	11	12	13	14	15	16
17	18	19	20	21	22	23
24	25	26	27	28	29	30

May

Su	Mo	Tu	We	Th	Fr	Sa
1	2	3	4	5	6	7
8	9	10	11	12	13	14
15	16	17	18	19	20	21
22	23	24	25	26	27	28
29	30	31	1	2	3	4

June

Su	Mo	Tu	We	Th	Fr
29	30	31	1	2	3
5	6	7	8	9	10
12	13	14	15	16	17
19	20	21	22	23	24
26	27	28	29	30	1

July

Su	Mo	Tu	We	Th	Fr	Sa
26	27	28	29	30	1	2
3	4	5	6	7	8	9
10	11	12	13	14	15	16
17	18	19	20	21	22	23
24	25	26	27	28	29	30
31	1	2	3	4	5	6

August

Su	Mo	Tu	We	Th	Fr	Sa
31	1	2	3	4	5	6
7	8	9	10	11	12	13
14	15	16	17	18	19	20
21	22	23	24	25	26	27
28	29	30	31	1	2	3

September

Su	Mo	Tu	We	Th	Fr
28	29	30	31	1	2
4	5	6	7	8	9
11	12	13	14	15	16
18	19	20	21	22	23
25	26	27	28	29	30

October

Su	Mo	Tu	We	Th	Fr	Sa
25	26	27	28	29	30	1
2	3	4	5	6	7	8
9	10	11	12	13	14	15
16	17	18	19	20	21	22
23	24	25	26	27	28	29
30	31	1	2	3	4	5

November

Su	Mo	Tu	We	Th	Fr	Sa
30	31	1	2	3	4	5
6	7	8	9	10	11	12
13	14	15	16	17	18	19
20	21	22	23	24	25	26
27	28	29	30	1	2	3

December

Su	Mo	Tu	We	Th	Fr
27	28	29	30	1	2
4	5	6	7	8	9
11	12	13	14	15	16
18	19	20	21	22	23
25	26	27	28	29	30

Annual Planner

January	February	March
April	May	June
July	August	September
October	November	December

Goals

Goals For This Year

Key
Objective _____

Goal Checklist

_____ □
_____ □
_____ □
_____ □
_____ □
_____ □
_____ □
_____ □
_____ □
_____ □
_____ □

Places to Visit

_____ □
_____ □
_____ □
_____ □
_____ □

People to Meet

_____ □
_____ □
_____ □
_____ □
_____ □

Notes

January 2022

Sun	Mon	Tue	Wed	Thu	Fri	Sat
26	27	28	29	30	31	1
2	3	4	5	6	7	8
9	10	11	12	13	14	15
16	17	18	19	20	21	22
23	24	25	26	27	28	29
30	31	1	2	3	4	5

December/January

Week 52

○ 27. MONDAY

PRIORITIES

_____
_____
_____

○ 28. TUESDAY

_____
_____
_____
_____

○ 29. WEDNESDAY

TO DO

_____

○ 30. THURSDAY

_____
_____
_____
_____
_____
_____

○ 31. FRIDAY

_____
_____
_____
_____
_____

○ 1. SATURDAY / 2. SUNDAY

_____
_____
_____
_____

January

01/03/22 - 01/09/22

---

3. MONDAY

PRIORITIES

4. TUESDAY

5. WEDNESDAY

TO DO

6. THURSDAY

7. FRIDAY

8. SATURDAY / 9. SUNDAY

January

Week 2

○ 10. MONDAY

PRIORITIES

○ 11. TUESDAY

○ 12. WEDNESDAY

TO DO

○ 13. THURSDAY

○ 14. FRIDAY

○ 15. SATURDAY / 16. SUNDAY

anuary

01/17/22 - 01/23/22

17. MONDAY

PRIORITIES

18. TUESDAY

19. WEDNESDAY

TO DO

20. THURSDAY

21. FRIDAY

22. SATURDAY / 23. SUNDAY

January

○ 24. MONDAY

PRIORITIES

○ 25. TUESDAY

○ 26. WEDNESDAY

TO DO

○ 27. THURSDAY

○ 28. FRIDAY

○ 29. SATURDAY / 30. SUNDAY

February 2022

Sun	Mon	Tue	Wed	Thu	Fri	Sat
30	31	1	2	3	4	5
6	7	8	9	10	11	12
13	14	15	16	17	18	19
20	21	22	23	24	25	26
27	28	1	2	3	4	5

January/February

01/31/22 - 02/06/22

○ 31. MONDAY

PRIORITIES

_____
_____
_____

○ 1. TUESDAY

_____
_____
_____
_____

○ 2. WEDNESDAY

TO DO

_____
_____
_____
_____

○ 3. THURSDAY

_____
_____
_____
_____

○ 4. FRIDAY

_____
_____
_____

○ 5. SATURDAY / 6. SUNDAY

_____
_____
_____
_____

ebruary

02/07/22 - 02/13/22

---

7. MONDAY

PRIORITIES

_____

_____

_____

8. TUESDAY

_____

_____

_____

_____

9. WEDNESDAY

TO DO

_____

_____

_____

10. THURSDAY

_____

_____

_____

_____

_____

11. FRIDAY

_____

_____

_____

_____

12. SATURDAY / 13. SUNDAY

_____

_____

_____

_____

February

○ 14. MONDAY

PRIORITIES

_____

_____

_____

○ 15. TUESDAY

_____

_____

_____

_____

○ 16. WEDNESDAY

TO DO

_____

_____

○ 17. THURSDAY

_____

_____

_____

_____

_____

○ 18. FRIDAY

_____

_____

_____

○ 19. SATURDAY / 20. SUNDAY

_____

_____

_____

February

21. MONDAY

PRIORITIES

_____

_____

_____

22. TUESDAY

_____

_____

_____

23. WEDNESDAY

TO DO

_____

_____

_____

24. THURSDAY

_____

_____

_____

_____

25. FRIDAY

_____

_____

_____

26. SATURDAY / 27. SUNDAY

_____

_____

_____

_____

March 2022

Sun	Mon	Tue	Wed	Thu	Fri	Sat
27	28	1	2	3	4	5
6	7	8	9	10	11	12
13	14	15	16	17	18	19
20	21	22	23	24	25	26
27	28	29	30	31	1	2

February/March

02/28/22 - 03/06/22

28. MONDAY

PRIORITIES

_____

_____

_____

1. TUESDAY

_____

_____

_____

_____

2. WEDNESDAY

TO DO

_____

_____

_____

3. THURSDAY

_____

_____

_____

_____

_____

4. FRIDAY

_____

_____

_____

_____

5. SATURDAY / 6. SUNDAY

_____

_____

_____

_____

March

Week 10

○ 7. MONDAY

PRIORITIES

○ 8. TUESDAY

○ 9. WEDNESDAY

TO DO

○ 10. THURSDAY

○ 11. FRIDAY

○ 12. SATURDAY / 13. SUNDAY

March

03/14/22 - 03/20/22

14. MONDAY

PRIORITIES

15. TUESDAY

16. WEDNESDAY

TO DO

17. THURSDAY

18. FRIDAY

19. SATURDAY / 20. SUNDAY

March

Week 12

○ 21. MONDAY

PRIORITIES

○ 22. TUESDAY

○ 23. WEDNESDAY

TO DO

○ 24. THURSDAY

○ 25. FRIDAY

○ 26. SATURDAY / 27. SUNDAY

April 2022

Sun	Mon	Tue	Wed	Thu	Fri	Sat
28	29	30	31	1	2	
4	5	6	7	8	9	
11	12	13	14	15	16	
18	19	20	21	22	23	
25	26	27	28	29	30	

March/April

Week 13

○ 28. MONDAY

PRIORITIES

○ 29. TUESDAY

○ 30. WEDNESDAY

TO DO

○ 31. THURSDAY

○ 1. FRIDAY

○ 2. SATURDAY / 3. SUNDAY

April

Week 14

4. MONDAY

PRIORITIES

5. TUESDAY

6. WEDNESDAY

TO DO

7. THURSDAY

8. FRIDAY

9. SATURDAY / 10. SUNDAY

April
Week 15

○ 11. MONDAY

PRIORITIES

_____

_____

○ 12. TUESDAY

_____

_____

_____

_____

○ 13. WEDNESDAY

TO DO

_____

_____

○ 14. THURSDAY

_____

_____

_____

_____

○ 15. FRIDAY

_____

_____

_____

○ 16. SATURDAY / 17. SUNDAY

_____

_____

_____

_____

04/18/22 - 04/24/22

18. MONDAY

PRIORITIES

19. TUESDAY

20. WEDNESDAY

TO DO

21. THURSDAY

22. FRIDAY

23. SATURDAY / 24. SUNDAY

May 2022

Sun	Mon	Tue	Wed	Thu	Fri	Sat
1	2	3	4	5	6	7
8	9	10	11	12	13	14
15	16	17	18	19	20	21
22	23	24	25	26	27	28
29	30	31	1	2	3	4

04/25/22 - 05/01/22

25. MONDAY

PRIORITIES

26. TUESDAY

27. WEDNESDAY

TO DO

28. THURSDAY

29. FRIDAY

30. SATURDAY / 1. SUNDAY

May
Week 18

○ 2. MONDAY

PRIORITIES

○ 3. TUESDAY

○ 4. WEDNESDAY

TO DO

○ 5. THURSDAY

○ 6. FRIDAY

○ 7. SATURDAY / 8. SUNDAY

May

05/09/22 - 05/15/22

9. MONDAY

PRIORITIES

10. TUESDAY

11. WEDNESDAY

TO DO

12. THURSDAY

13. FRIDAY

14. SATURDAY / 15. SUNDAY

May

Week 20

○ 16. MONDAY

PRIORITIES

○ 17. TUESDAY

○ 18. WEDNESDAY

TO DO

○ 19. THURSDAY

○ 20. FRIDAY

○ 21. SATURDAY / 22. SUNDAY

May

Week 21

23. MONDAY

PRIORITIES

24. TUESDAY

25. WEDNESDAY

TO DO

26. THURSDAY

27. FRIDAY

28. SATURDAY / 29. SUNDAY

June 2022

Sun	Mon	Tue	Wed	Thu	Fri	Sat
29	30	31	1	2	3	4
5	6	7	8	9	10	11
12	13	14	15	16	17	18
19	20	21	22	23	24	25
26	27	28	29	30	1	2

May/June

05/30/22 - 06/05/22

30. MONDAY

PRIORITIES

31. TUESDAY

1. WEDNESDAY

TO DO

2. THURSDAY

3. FRIDAY

4. SATURDAY / 5. SUNDAY

June

○ 6. MONDAY

PRIORITIES

○ 7. TUESDAY

○ 8. WEDNESDAY

TO DO

○ 9. THURSDAY

○ 10. FRIDAY

○ 11. SATURDAY / 12. SUNDAY

June

13. MONDAY

PRIORITIES

_____
_____
_____

14. TUESDAY

_____
_____
_____
_____

15. WEDNESDAY

TO DO

_____
_____
_____

16. THURSDAY

_____
_____
_____
_____
_____
_____

17. FRIDAY

_____
_____
_____
_____

18. SATURDAY / 19. SUNDAY

_____
_____
_____
_____
_____

June

Week 25

○ 20. MONDAY

PRIORITIES

○ 21. TUESDAY

○ 22. WEDNESDAY

TO DO

○ 23. THURSDAY

○ 24. FRIDAY

○ 25. SATURDAY / 26. SUNDAY

July 2022

Sun	Mon	Tue	Wed	Thu	Fri	Sat
27	28	29	30	1	2	
4	5	6	7	8	9	
11	12	13	14	15	16	
18	19	20	21	22	23	
25	26	27	28	29	30	
1	2	3	4	5	6	

June/July

Week 26

○ 27. MONDAY

PRIORITIES

_____
_____
_____

○ 28. TUESDAY

_____
_____
_____
_____

○ 29. WEDNESDAY

TO DO

_____
_____

○ 30. THURSDAY

_____
_____
_____
_____
_____

○ 1. FRIDAY

_____
_____

○ 2. SATURDAY / 3. SUNDAY

_____
_____
_____
_____

July

07/04/22 - 07/10/22

4. MONDAY

PRIORITIES

5. TUESDAY

6. WEDNESDAY

TO DO

7. THURSDAY

8. FRIDAY

9. SATURDAY / 10. SUNDAY

July
Week 28

○ 11. MONDAY

PRIORITIES

_____

_____

_____

○ 12. TUESDAY

_____

_____

_____

_____

○ 13. WEDNESDAY

TO DO

_____

_____

_____

○ 14. THURSDAY

_____

_____

_____

_____

_____

○ 15. FRIDAY

_____

_____

_____

○ 16. SATURDAY / 17. SUNDAY

_____

_____

_____

_____

uly

07/18/22 - 07/24/22

18. MONDAY

PRIORITIES

19. TUESDAY

20. WEDNESDAY

TO DO

21. THURSDAY

22. FRIDAY

23. SATURDAY / 24. SUNDAY

July
Week 30

○ 25. MONDAY

PRIORITIES

○ 26. TUESDAY

○ 27. WEDNESDAY

TO DO

○ 28. THURSDAY

○ 29. FRIDAY

○ 30. SATURDAY / 31. SUNDAY

August 2022

Sun	Mon	Tue	Wed	Thu	Fri	Sat
31	1	2	3	4	5	6
7	8	9	10	11	12	13
14	15	16	17	18	19	20
21	22	23	24	25	26	27
28	29	30	31	1	2	3

August
Week 31

○ 1. MONDAY

PRIORITIES

○ 2. TUESDAY

○ 3. WEDNESDAY

TO DO

○ 4. THURSDAY

○ 5. FRIDAY

○ 6. SATURDAY / 7. SUNDAY

08/08/22 - 08/14/22

8. MONDAY

PRIORITIES

_____
_____
_____
_____
_____

9. TUESDAY

_____
_____

10. WEDNESDAY

TO DO

_____
_____
_____

11. THURSDAY

_____
_____
_____
_____
_____

12. FRIDAY

_____
_____
_____
_____

13. SATURDAY / 14. SUNDAY

_____
_____
_____
_____

August
Week 33

○ 15. MONDAY

PRIORITIES

○ 16. TUESDAY

○ 17. WEDNESDAY

TO DO

○ 18. THURSDAY

○ 19. FRIDAY

○ 20. SATURDAY / 21. SUNDAY

August

22. MONDAY

PRIORITIES

23. TUESDAY

24. WEDNESDAY

TO DO

25. THURSDAY

26. FRIDAY

27. SATURDAY / 28. SUNDAY

September 2022

Sun	Mon	Tue	Wed	Thu	Fri	Sat
28	29	30	31	1	2	3
4	5	6	7	8	9	10
11	12	13	14	15	16	17
18	19	20	21	22	23	24
25	26	27	28	29	30	1

ugust/September

08/29/22 - 09/04/22

29. MONDAY

PRIORITIES

30. TUESDAY

31. WEDNESDAY

TO DO

1. THURSDAY

2. FRIDAY

3. SATURDAY / 4. SUNDAY

September
Week 36

○ 5. MONDAY

PRIORITIES

○ 6. TUESDAY

○ 7. WEDNESDAY

TO DO

○ 8. THURSDAY

○ 9. FRIDAY

○ 10. SATURDAY / 11. SUNDAY

September

09/12/22 - 09/18/22

---

12. MONDAY

PRIORITIES

_____
_____
_____
_____

13. TUESDAY

_____
_____
_____

14. WEDNESDAY

TO DO

_____
_____
_____

15. THURSDAY

_____
_____
_____
_____

16. FRIDAY

_____
_____
_____

17. SATURDAY / 18. SUNDAY

_____
_____
_____

September

Week 38

○ 19. MONDAY

PRIORITIES

○ 20. TUESDAY

○ 21. WEDNESDAY

TO DO

○ 22. THURSDAY

○ 23. FRIDAY

○ 24. SATURDAY / 25. SUNDAY

October 2022

Sun	Mon	Tue	Wed	Thu	Fri	Sat
26	27	28	29	30	1	
3	4	5	6	7	8	
10	11	12	13	14	15	
17	18	19	20	21	22	
24	25	26	27	28	29	
31	1	2	3	4	5	

September/October

Week 39

○ 26. MONDAY

PRIORITIES

_____
_____
_____
○ 27. TUESDAY _____
_____
_____
_____
_____

○ 28. WEDNESDAY

TO DO

_____
_____
_____
○ 29. THURSDAY _____
_____
_____
_____
_____
_____
○ 30. FRIDAY _____
_____
_____
_____
○ 1. SATURDAY / 2. SUNDAY _____
_____
_____
_____
_____

October

10/03/22 - 10/09/22

3. MONDAY

PRIORITIES

4. TUESDAY

5. WEDNESDAY

TO DO

6. THURSDAY

7. FRIDAY

8. SATURDAY / 9. SUNDAY

October

Week 41

○ 10. MONDAY

PRIORITIES

○ 11. TUESDAY

○ 12. WEDNESDAY

TO DO

○ 13. THURSDAY

○ 14. FRIDAY

○ 15. SATURDAY / 16. SUNDAY

October

10/17/22 - 10/23/22

17. MONDAY

PRIORITIES

_____

_____

_____

18. TUESDAY

_____

_____

_____

_____

19. WEDNESDAY

TO DO

_____

_____

_____

20. THURSDAY

_____

_____

_____

_____

_____

21. FRIDAY

_____

_____

_____

22. SATURDAY / 23. SUNDAY

_____

_____

_____

_____

October

Week 43

○ 24. MONDAY

PRIORITIES

○ 25. TUESDAY

○ 26. WEDNESDAY

TO DO

○ 27. THURSDAY

○ 28. FRIDAY

○ 29. SATURDAY / 30. SUNDAY

November 2022

Sun	Mon	Tue	Wed	Thu	Fri	Sat
30	31	1	2	3	4	5
6	7	8	9	10	11	12
13	14	15	16	17	18	19
20	21	22	23	24	25	26
27	28	29	30	1	2	3

October/November

○ 31. MONDAY

PRIORITIES

○ 1. TUESDAY

○ 2. WEDNESDAY

TO DO

○ 3. THURSDAY

○ 4. FRIDAY

○ 5. SATURDAY / 6. SUNDAY

November

11/07/22 - 11/13/22

7. MONDAY

PRIORITIES

8. TUESDAY

9. WEDNESDAY

TO DO

10. THURSDAY

11. FRIDAY

12. SATURDAY / 13. SUNDAY

November

Week 46

○ 14. MONDAY

PRIORITIES

○ 15. TUESDAY

○ 16. WEDNESDAY

TO DO

○ 17. THURSDAY

○ 18. FRIDAY

○ 19. SATURDAY / 20. SUNDAY

November

11/21/22 - 11/27/22

21. MONDAY

PRIORITIES

22. TUESDAY

23. WEDNESDAY

TO DO

24. THURSDAY

25. FRIDAY

26. SATURDAY / 27. SUNDAY

December 2022

Sun	Mon	Tue	Wed	Thu	Fri	Sat
27	28	29	30	1	2	3
4	5	6	7	8	9	10
11	12	13	14	15	16	17
18	19	20	21	22	23	24
25	26	27	28	29	30	31

11/28/22 - 12/04/22

28. MONDAY

PRIORITIES

29. TUESDAY

30. WEDNESDAY

TO DO

1. THURSDAY

2. FRIDAY

3. SATURDAY / 4. SUNDAY

December

Week 49

○ 5. MONDAY

PRIORITIES

_____

_____

_____

○ 6. TUESDAY

_____

_____

_____

_____

○ 7. WEDNESDAY

TO DO

_____

_____

_____

○ 8. THURSDAY

_____

_____

_____

_____

_____

○ 9. FRIDAY

_____

_____

_____

_____

○ 10. SATURDAY / 11. SUNDAY

_____

_____

_____

_____

December

12. MONDAY

PRIORITIES
_____
_____
_____

13. TUESDAY
_____
_____
_____
_____

14. WEDNESDAY

TO DO
_____
_____
_____

15. THURSDAY
_____
_____
_____
_____
_____

16. FRIDAY
_____
_____
_____
_____

17. SATURDAY / 18. SUNDAY
_____
_____
_____
_____

December

Week 51

12/19/22 - 12/25/22

○ 19. MONDAY

PRIORITIES

_____

_____

○ 20. TUESDAY

_____

_____

_____

_____

○ 21. WEDNESDAY

TO DO

_____

○ 22. THURSDAY

_____

_____

_____

_____

○ 23. FRIDAY

_____

_____

_____

○ 24. SATURDAY / 25. SUNDAY

_____

_____

_____

_____

December

12/26/22 - 01/01/23

---

26. MONDAY

PRIORITIES

27. TUESDAY

28. WEDNESDAY

TO DO

29. THURSDAY

30. FRIDAY

31. SATURDAY / 1. SUNDAY

Subscription Tracker

Organisation	Date Paid	Amount	Duration	Exp. Date	Method of Renewal	Renewal Date

Subscription Tracker

Organisation	Date Paid	Amount	Duration	Exp. Date	Method of Renewal	Renewal Date

Passwords

Website	
Username	
Password	
Email	
Notes	

Website	
Username	
Password	
Email	
Notes	

Website	
Username	
Password	
Email	
Notes	

Website	
Username	
Password	
Email	
Notes	

Website	
Username	
Password	
Email	
Notes	

Website	
Username	
Password	
Email	
Notes	

Website	
Username	
Password	
Email	
Notes	

Website	
Username	
Password	
Email	
Notes	

Passwords

Website	
Username	
Password	
Email	
Notes	

Website	
Username	
Password	
Email	
Notes	

Website	
Username	
Password	
Email	
Notes	

Website	
Username	
Password	
Email	
Notes	

Website	
Username	
Password	
Email	
Notes	

Website	
Username	
Password	
Email	
Notes	

Website	
Username	
Password	
Email	
Notes	

Website	
Username	
Password	
Email	
Notes	

Passwords

Website		Website	
Username		Username	
Password		Password	
Email		Email	
Notes		Notes	

Website		Website	
Username		Username	
Password		Password	
Email		Email	
Notes		Notes	

Website		Website	
Username		Username	
Password		Password	
Email		Email	
Notes		Notes	

Website		Website	
Username		Username	
Password		Password	
Email		Email	
Notes		Notes	

Contacts

Name

Address

City   State   Zip

Phone

Email

Name

Address

City   State   Zip

Phone

Email

Name

Address

City   State   Zip

Phone

Email

Name

Address

City   State   Zip

Phone

Email

Name

Address

City   State   Zip

Phone

Email

Name

Address

City   State   Zip

Phone

Email

Name

Address

City   State   Zip

Phone

Email

Name

Address

City   State   Zip

Phone

Email

Name

Address

City   State   Zip

Phone

Email

Name

Address

City   State   Zip

Phone

Email

Contacts

Name

Address

City   State   Zip

Phone

Email

Name

Address

City   State   Zip

Phone

Email

Name

Address

City   State   Zip

Phone

Email

Name

Address

City   State   Zip

Phone

Email

Name

Address

City   State   Zip

Phone

Email

Name

Address

City   State   Zip

Phone

Email

Name

Address

City   State   Zip

Phone

Email

Name

Address

City   State   Zip

Phone

Email

Name

Address

City   State   Zip

Phone

Email

Name

Address

City   State   Zip

Phone

Email

Contacts

ne		
ress		
	State	Zip
ne		
il		

Name

Address

City   State   Zip

Phone

Email

ne		
ress		
	State	Zip
ne		
il		

Name

Address

City   State   Zip

Phone

Email

ne		
ress		
	State	Zip
ne		
il		

Name

Address

City   State   Zip

Phone

Email

ne		
ress		
	State	Zip
ne		
il		

Name

Address

City   State   Zip

Phone

Email

ne		
ress		
	State	Zip
ne		
il		

Name

Address

City   State   Zip

Phone

Email

Birthday Reminder

JANUARY	FEBRUARY	MARCH

APRIL	MAY	JUNE

JULY	AUGUST	SEPTEMBER

OCTOBER	NOVEMBER	DECEMBER

My Favorite Movies

Title:_____
Actors:_____
_____
_____
Published:_____
Genre:_____
Ranking: ☆ ☆ ☆ ☆ ☆

Title:_____
Actors:_____
_____
_____
Published:_____
Genre:_____
Ranking: ☆ ☆ ☆ ☆ ☆

Title:_____
Actors:_____
_____
_____
Published:_____
Genre:_____
Ranking: ☆ ☆ ☆ ☆ ☆

Title:_____
Actors:_____
_____
_____
Published:_____
Genre:_____
Ranking: ☆ ☆ ☆ ☆ ☆

Title:_____
Actors:_____
_____
_____
Published:_____
Genre:_____
Ranking: ☆ ☆ ☆ ☆ ☆

Title:_____
Actors:_____
_____
_____
Published:_____
Genre:_____
Ranking: ☆ ☆ ☆ ☆ ☆

Title:_____
Actors:_____
_____
_____
Published:_____
Genre:_____
Ranking: ☆ ☆ ☆ ☆ ☆

Title:_____
Actors:_____
_____
_____
Published:_____
Genre:_____
Ranking: ☆ ☆ ☆ ☆ ☆

Title:_____
Actors:_____
_____
_____
Published:_____
Genre:_____
Ranking: ☆ ☆ ☆ ☆ ☆

Title:_____
Actors:_____
_____
_____
Published:_____
Genre:_____
Ranking: ☆ ☆ ☆ ☆ ☆

Title:_____
Actors:_____
_____
_____
Published:_____
Genre:_____
Ranking: ☆ ☆ ☆ ☆ ☆

Title:_____
Actors:_____
_____
_____
Published:_____
Genre:_____
Ranking: ☆ ☆ ☆ ☆ ☆

My Favorite Movies

Title:_____
Actors:_____
_____
_____
Published:_____
Genre:_____
Ranking: ☆ ☆ ☆ ☆ ☆

Title:_____
Actors:_____
_____
_____
Published:_____
Genre:_____
Ranking: ☆ ☆ ☆ ☆ ☆

Title:_____
Actors:_____
_____
_____
Published:_____
Genre:_____
Ranking: ☆ ☆ ☆ ☆ ☆

Title:_____
Actors:_____
_____
_____
Published:_____
Genre:_____
Ranking: ☆ ☆ ☆ ☆ ☆

Title:_____
Actors:_____
_____
_____
Published:_____
Genre:_____
Ranking: ☆ ☆ ☆ ☆ ☆

Title:_____
Actors:_____
_____
_____
Published:_____
Genre:_____
Ranking: ☆ ☆ ☆ ☆ ☆

Title:_____
Actors:_____
_____
_____
Published:_____
Genre:_____
Ranking: ☆ ☆ ☆ ☆ ☆

Title:_____
Actors:_____
_____
_____
Published:_____
Genre:_____
Ranking: ☆ ☆ ☆ ☆ ☆

Title:_____
Actors:_____
_____
_____
Published:_____
Genre:_____
Ranking: ☆ ☆ ☆ ☆ ☆

Title:_____
Actors:_____
_____
_____
Published:_____
Genre:_____
Ranking: ☆ ☆ ☆ ☆ ☆

Title:_____
Actors:_____
_____
_____
Published:_____
Genre:_____
Ranking: ☆ ☆ ☆ ☆ ☆

Title:_____
Actors:_____
_____
_____
Published:_____
Genre:_____
Ranking: ☆ ☆ ☆ ☆ ☆

Podcast List

dcast	Genre	Rating
		☆☆☆☆☆
		☆☆☆☆☆
		☆☆☆☆☆
		☆☆☆☆☆
		☆☆☆☆☆
		☆☆☆☆☆
		☆☆☆☆☆
		☆☆☆☆☆
		☆☆☆☆☆
		☆☆☆☆☆
		☆☆☆☆☆
		☆☆☆☆☆
		☆☆☆☆☆
		☆☆☆☆☆
		☆☆☆☆☆
		☆☆☆☆☆
		☆☆☆☆☆
		☆☆☆☆☆
		☆☆☆☆☆
		☆☆☆☆☆
		☆☆☆☆☆
		☆☆☆☆☆
		☆☆☆☆☆
		☆☆☆☆☆

Podcast List

Podcast	Genre	Rating
		☆☆☆☆☆
		☆☆☆☆☆
		☆☆☆☆☆
		☆☆☆☆☆
		☆☆☆☆☆
		☆☆☆☆☆
		☆☆☆☆☆
		☆☆☆☆☆
		☆☆☆☆☆
		☆☆☆☆☆
		☆☆☆☆☆
		☆☆☆☆☆
		☆☆☆☆☆
		☆☆☆☆☆
		☆☆☆☆☆
		☆☆☆☆☆
		☆☆☆☆☆
		☆☆☆☆☆
		☆☆☆☆☆
		☆☆☆☆☆
		☆☆☆☆☆
		☆☆☆☆☆
		☆☆☆☆☆
		☆☆☆☆☆

Wishlist

WHAT	PRICE

Gift Ideas For Family and Friends

NAME	IDEA	PRICE

Notes

Notes

Notes

Notes

Notes

Notes

Notes

Notes

Notes

Notes

Doodle Page

Doodle Page

Doodle Page

Doodle Page

Doodle Page

2023

January

Su	Mo	Tu	We	Th	Fr	Sa
1	2	3	4	5	6	7
8	9	10	11	12	13	14
15	16	17	18	19	20	21
22	23	24	25	26	27	28
29	30	31	1	2	3	4

February

Su	Mo	Tu	We	Th	Fr	Sa
29	30	31	1	2	3	4
5	6	7	8	9	10	11
12	13	14	15	16	17	18
19	20	21	22	23	24	25
26	27	28	1	2	3	4

March

Su	Mo	Tu	We	Th	Fr	S
26	27	28	1	2	3	
5	6	7	8	9	10	1
12	13	14	15	16	17	1
19	20	21	22	23	24	2
26	27	28	29	30	31	

April

Su	Mo	Tu	We	Th	Fr	Sa
26	27	28	29	30	31	1
2	3	4	5	6	7	8
9	10	11	12	13	14	15
16	17	18	19	20	21	22
23	24	25	26	27	28	29
30	1	2	3	4	5	6

May

Su	Mo	Tu	We	Th	Fr	Sa
30	1	2	3	4	5	6
7	8	9	10	11	12	13
14	15	16	17	18	19	20
21	22	23	24	25	26	27
28	29	30	31	1	2	3

June

Su	Mo	Tu	We	Th	Fr	S
28	29	30	31	1	2	
4	5	6	7	8	9	1
11	12	13	14	15	16	1
18	19	20	21	22	23	2
25	26	27	28	29	30	

July

Su	Mo	Tu	We	Th	Fr	Sa
25	26	27	28	29	30	1
2	3	4	5	6	7	8
9	10	11	12	13	14	15
16	17	18	19	20	21	22
23	24	25	26	27	28	29
30	31	1	2	3	4	5

August

Su	Mo	Tu	We	Th	Fr	Sa
30	31	1	2	3	4	5
6	7	8	9	10	11	12
13	14	15	16	17	18	19
20	21	22	23	24	25	26
27	28	29	30	31	1	2

September

Su	Mo	Tu	We	Th	Fr	S
27	28	29	30	31	1	
3	4	5	6	7	8	9
10	11	12	13	14	15	1
17	18	19	20	21	22	2
24	25	26	27	28	29	30

October

Su	Mo	Tu	We	Th	Fr	Sa
1	2	3	4	5	6	7
8	9	10	11	12	13	14
15	16	17	18	19	20	21
22	23	24	25	26	27	28
29	30	31	1	2	3	4

November

Su	Mo	Tu	We	Th	Fr	Sa
29	30	31	1	2	3	4
5	6	7	8	9	10	11
12	13	14	15	16	17	18
19	20	21	22	23	24	25
26	27	28	29	30	1	2

December

Su	Mo	Tu	We	Th	Fr	S
26	27	28	29	30	1	2
3	4	5	6	7	8	9
10	11	12	13	14	15	1
17	18	19	20	21	22	2
24	25	26	27	28	29	30
31	1	2	3	4	5	6

Made in the USA
Monee, IL
30 March 2022

93848515R00057